ROCKY'S AWESOME ADVENTURE

by Peggy Price

ISBN-13: 978-1505205107
ISBN-10: 1505205107

Design ©2015 Laurie Barrows
"Making the World a Happier Place, One Smile at a Time" ™

www.LaurieBarrows.com

Printed in the United States of America

Published in the United States of America

To Bryden, who rocks!

May he never lose his delight in the world around him, nor his curiosity to learn about it.

Sometimes it seems that things take a very, very long time to happen. HOW can it take soooo long for a birthday to come or to earn enough money to buy a bike?

But time is a funny thing. Every kind of creature has its own unique way of seeing time: a butterfly who might only live a month, a tree that could live to 500 years, ...or a rock.

In this story we're going to look at time in the way a ROCK sees it and learn that Life is a Dance that circles, every-changing, never-ending, always amazing.

This is an exciting adventure story that started "a long, LOOONG time ago". And every time I say that ("long, looong time ago"), you can say it, too! So listen well.

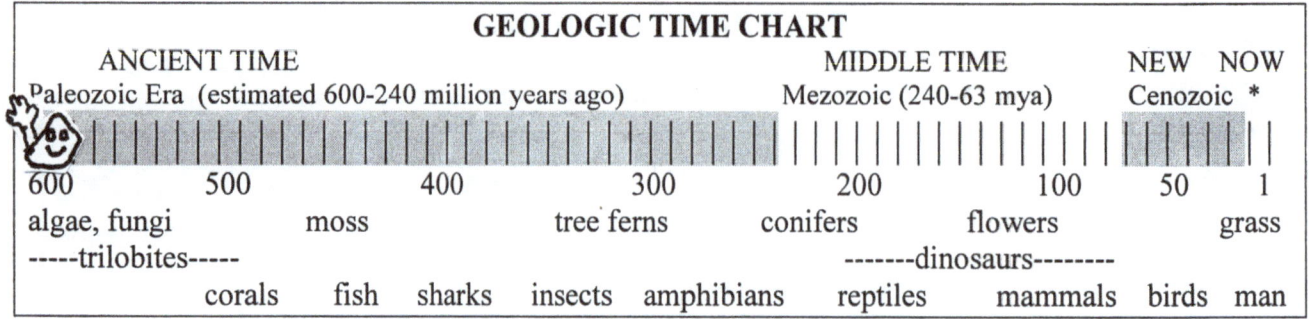

GEOLOGIC TIME CHART

ANCIENT TIME						MIDDLE TIME		NEW NOW

Paleozoic Era (estimated 600-240 million years ago)   Mezozoic (240-63 mya)   Cenozoic *

600	500	400	300	200	100	50	1
algae, fungi	moss	tree ferns	conifers		flowers		grass
-----trilobites-----				-------dinosaurs--------			
	corals	fish	sharks	insects	amphibians	reptiles	mammals	birds	man

Close your eyes and pretend it's six hundred million years ago...
that's a long LOOONG time ago....how long ago was it?(...)

The world is already very, VERY old but not much is living yet
on the still-forming land of the earth and only algae, bacteria
and very simple, soft-bodied aquatic animals live in the sea.

Jellyfish (early muscles)

Worm (early backbone)

2

Mother Nature can be very creative, however, and "suddenly" (over millions of years!), bacteria started morphing (changing) into a wild assortment of animals. There were worms with the start of brains and backbones and swirling jellyfish with the first muscles.

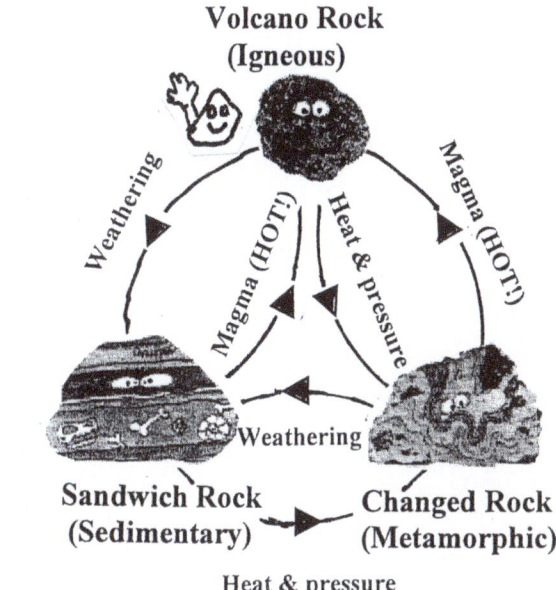

Volcano Rock
(Igneous)

Weathering

Magma (HOT!)

Heat & pressure

Magma (HOT!)

Weathering

Sandwich Rock
(Sedimentary)

Changed Rock
(Metamorphic)

Heat & pressure

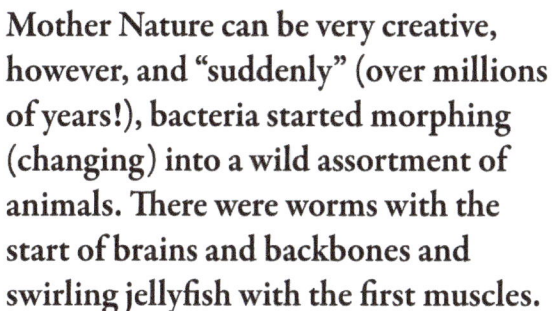

Ammonite

Trilobite (first eyes)

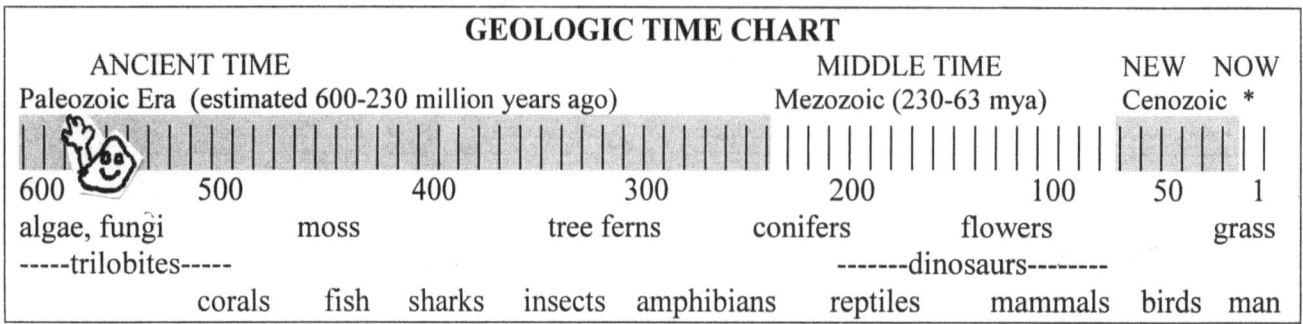

GEOLOGIC TIME CHART

ANCIENT TIME		MIDDLE TIME	NEW	NOW

Paleozoic Era (estimated 600-230 million years ago)　　　Mezozoic (230-63 mya)　　Cenozoic *

600　　　　500　　　　　400　　　　　300　　　　　200　　　　　100　　　50　　　1

algae, fungi　　　　moss　　　　　tree ferns　　　conifers　　　　flowers　　　　　grass

-----trilobites-----　　　　　　　　　　　　　　　　-------dinosaurs--------

　　　　corals　fish　sharks　insects　amphibians　reptiles　　mammals　birds　man

And there were hard-shelled trilobites
(with the very first eyes).

The trilobite in this picture was pressed
into the sand of an ocean bed millions
of years ago and made into a
rock-like fossil.

Fossils of long-ago animals and plants
are some of the clues that make this
mystery story possible.

They help us to understand
and read the story of rocks.

Penn Dixie Fossils collected from the Windom Shale by Dan Cooper

4

In this long-ago storm-torn world a
Met-a-mor-phic rock was sitting
peacefully on the top of a mountain.

We'll call him Rocky.

Can you find Rocky (as a
Met-a-mor-phic or "changed" rock)
on the rock cycle circle here?

All of a sudden the rain and the wind
broke him lose, tumbling and falling
down the mountain side...crash,
bang, thunk.

"Help, help!" he cries.

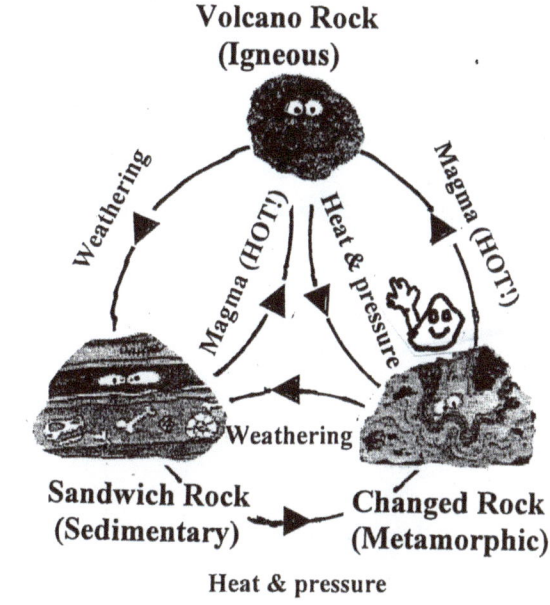

**Volcano Rock
(Igneous)**

Weathering

Magma (HOT!)

Heat & pressure

Magma (HOT!)

Weathering

**Sandwich Rock
(Sedimentary)**

**Changed Rock
(Metamorphic)**

Heat & pressure

What do you think will happen to Rocky?

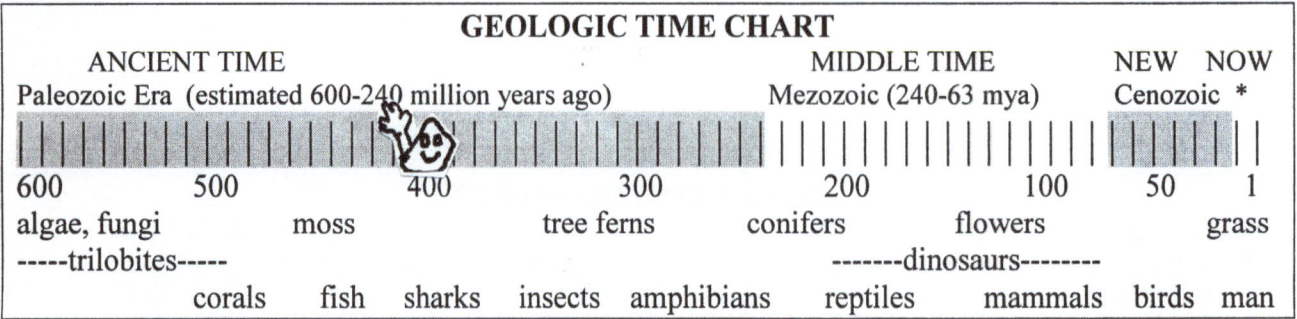

GEOLOGIC TIME CHART

ANCIENT TIME				MIDDLE TIME		NEW	NOW
Paleozoic Era (estimated 600-240 million years ago)				Mezozoic (240-63 mya)		Cenozoic	*

600	500	400	300	200	100	50	1
algae, fungi	moss		tree ferns	conifers	flowers		grass
-----trilobites-----				-------dinosaurs--------			
	corals	fish	sharks	insects	amphibians	reptiles	mammals	birds	man

Two hundred million years have passed. It's now 400 million years ago. Can you find that on the chart? That's a long LOONG time ago, isn't it?!(...) But NOW there are sharks in the sea. The algae have morphed into fungi (like mushrooms), which "eat" soil, and into real plants (like ferns) which "eat" sunshine to grow. They are starting to cover the land.

After his fall, our friend Rocky has been pushed down, down, DOWN into the center of the earth where it is very, VERY hot & Rocky is now melted into red-hot lava.

SUDDENLY... the earth shifts and with a huge BANG! Rocky ERUPTS up through a volcano back onto the top of the earth.

"Wheee, what a ride!"
Red hot, Rocky flows down the mountain until finally he cools... aaah!

"I can see the sky again!" Rocky thinks.

Lava

Volcano

Magma

"Hi, I'm Rocky," he says shyly to the new rocks that are now around him.

As fun as that ride was, it was a little scary, too and now he has to make friends again.

"Are you ever lonely?" he asks Ned, a Met-a-mor-phick rock (remember that's a CHANGED rock) next to him.

"No, not usually," answers Ned, "Because we're ALL together on the roller coaster ROCK CYCLE. No matter how we change, we're all together."

"But it's so scary," Rocky mumbles.

"Yes, it can be scary but it can also be exciting and interesting, if you let it be," replies Ned.

"What can possibly happen to me next? I just know it's gonna be scary!" thinks Rocky.

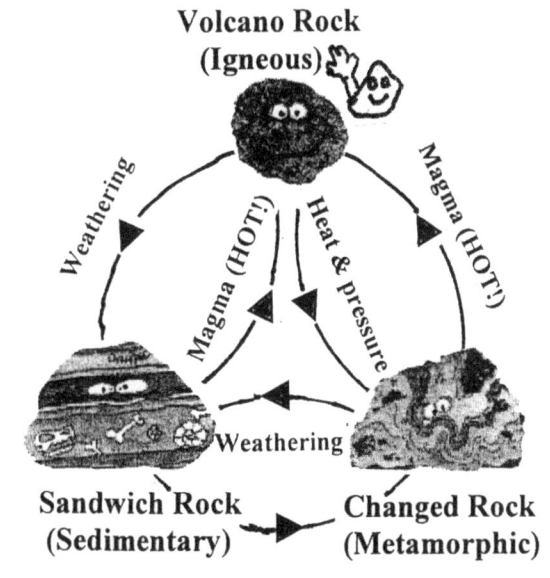

Volcano Rock (Igneous)

Weathering

Magma (HOT!)

Heat & pressure

Magma (HOT!)

Weathering

Sandwich Rock (Sedimentary)

Changed Rock (Metamorphic)

Heat & pressure

Coming from a volcano, Rocky is now an Igneous rock, a VOLCANO-ROCK.

Can you say IGNEOUS (ig-nee-us)?

Can you find Rocky in his circle of travels?

Rocks like Obsidian and Granite and Pumice are Igneous.

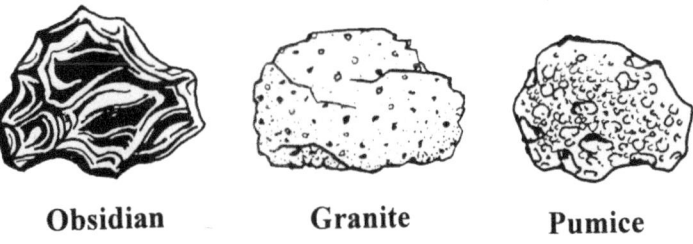

Obsidian   Granite   Pumice

7

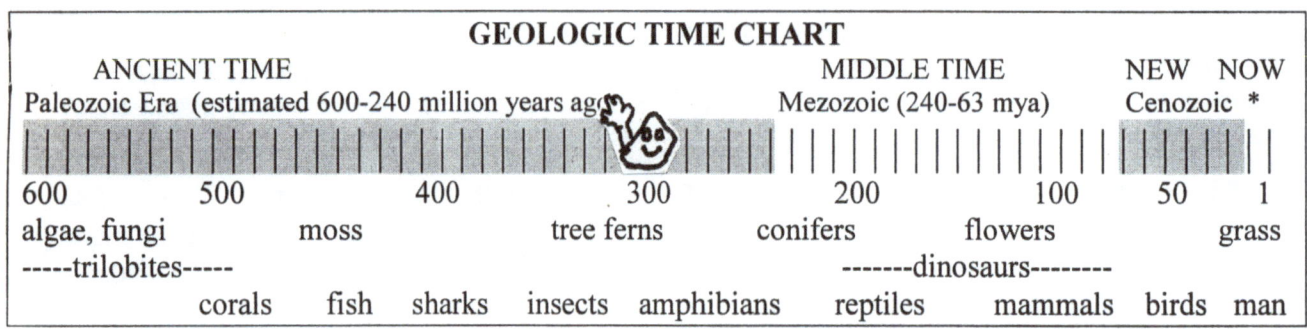

GEOLOGIC TIME CHART

ANCIENT TIME		MIDDLE TIME	NEW	NOW
Paleozoic Era (estimated 600-240 million years ago)		Mezozoic (240-63 mya)	Cenozoic	*

600	500	400	300	200	100	50	1
algae, fungi	moss		tree ferns	conifers	flowers		grass

-----trilobites-----　　　　　　　　　　　　　　　　　-------dinosaurs--------

　　　　　corals　　fish　　sharks　　insects　amphibians　　reptiles　　mammals　birds　man

ANOTHER hundred million years have gone by. It's now 300 million years ago. Where is that on our time chart? So this is still a long, LOOONG time ago (....) But NOW much of the world is a SAMP. Amphibians RULE! What's an amphibian? (an animal with a smooth skin that lives partly in the water and partly on land).

But there are also HUGE insects living in this swamp with its tree-sized ferns: cockroaches up to 4 inches long! and dragonflies with wings over two feet wide...how big is that? They were the biggest insects of all time. The air is alive with their buzz. Can you imagine them zooming by like mini jets with huge eyes?

Life as an Ig-ne-ous VOLCANO rock has been good for Rocky. He's seen a lot. But time and weather are wearing Rocky away... gradually, one tiny piece at a time into small pebbles. All those little pieces, like grains of sand, are part of what makes up the earth's soil.

Volcano Rock
(Igneous)

Weathering

Magma (HOT!)

Heat & pressure

Magma (HOT!)

Weathering

Sandwich Rock
(Sedimentary)

Changed Rock
(Metamorphic)

Heat & pressure

Can you find in the circle where the Volcano Rocky is being weathered away?

A smaller Rocky is rolling along in larger and larger rivers until he finally comes to the sea.

Do YOU think that's the end of Rocky?

Wait and see.

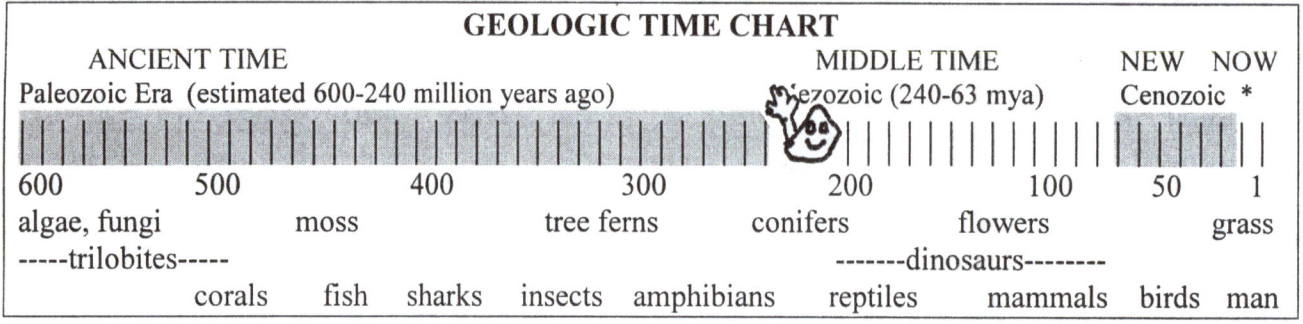

GEOLOGIC TIME CHART

ANCIENT TIME	MIDDLE TIME	NEW NOW
Paleozoic Era (estimated 600-240 million years ago)	Mezozoic (240-63 mya)	Cenozoic *

600	500	400	300	200	100	50	1

algae, fungi   moss   tree ferns   conifers   flowers   grass

-----trilobites-----   -------dinosaurs--------

corals   fish   sharks   insects   amphibians   reptiles   mammals   birds   man

Now it's 225 million years ago. Can you find that on our time chart? Is that a long, LOOONG time ago? (...) Warm salt water seas cover much of the earth with coastal swamps and dry, dusty deserts inland.

It's the Age of REPTILES. Crocodiles, lizards, turtles and snakes are here already (my goodness, but they've been on the earth for a very long time!).

Palm-like cycad plants and towering redwood-like trees cover the earth while huge Thec-o-dont lizards (early Dinosaurs) hunt voraciously for insects in the sunny summer-year-round weather.

Rocky (remember, he's only grains of sand now) is at the bottom of the ocean with sea worms burrowing near him. Rocky becomes hotter and hotter as he is buried deeper under more and more sediment. Pressed together with lots of other grains of sand (and sometimes insects, plants and fish that have died and will become fossils), he is cemented into a brand new rock as the ocean dries up.

Where is Rocky on his roller coaster Rock Cycle ride?
Do you think THAT'S the end of Rocky's story?

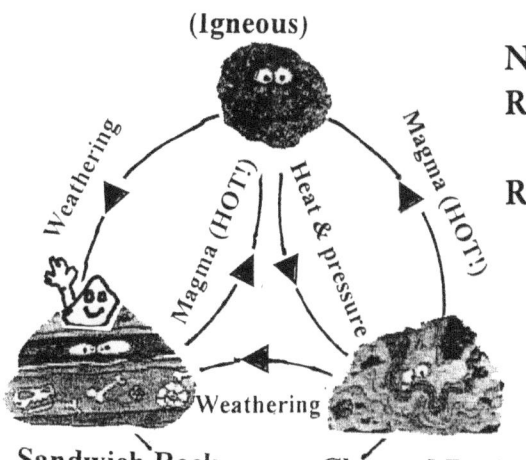

Now Rocky is a Sed-i-men-ta-ry rock...a squished SANDWICH ROCK. He looks very different than he used to look.

Rocks like Sandstone, Conglomerate and Limestone are sedimentary.

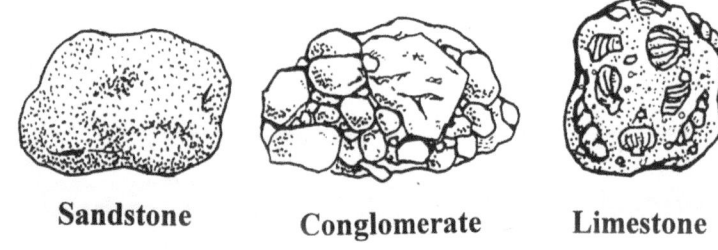

"Hi, I'm Rocky," he says to the other sedimentary rocks next to him.

"HI," they all reply.

"It's hard having to make new friends," he whispers to the rock closest to him.

"But I'm your friend," says the rock. I knew you when you were an Ig-ne-ous VOLCANO Rock. You look different now but I'm still your friend. You have LOTS of friends."

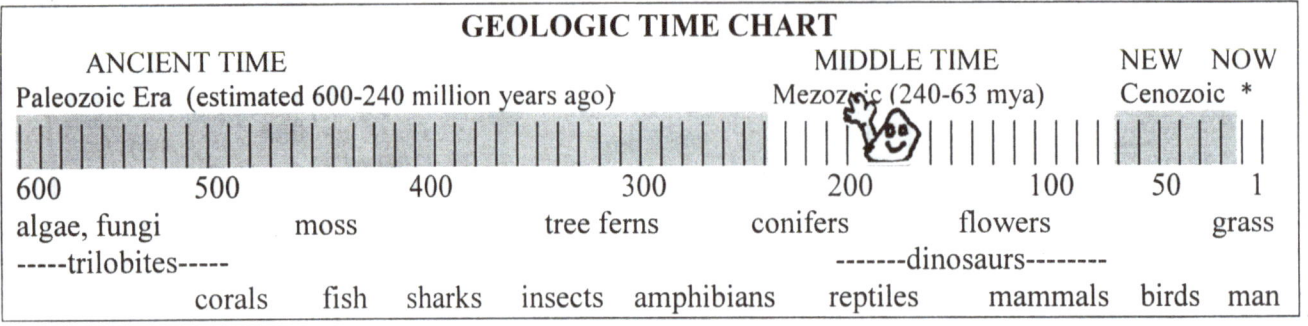

GEOLOGIC TIME CHART

ANCIENT TIME				MIDDLE TIME		NEW NOW
Paleozoic Era (estimated 600-240 million years ago)				Mezozoic (240-63 mya)		Cenozoic *

600	500	400	300	200	100	50	1
algae, fungi	moss		tree ferns	conifers	flowers		grass
-----trilobites-----					-------dinosaurs--------		
	corals	fish	sharks	insects	amphibians	reptiles	mammals	birds	man

It's 180 million years ago, the Jurassic age. That's …a long LOOONG time ago. (…) Dinosaurs roar and wander the earth. There are fearsome Tyrannosaurs and gentle plant-eating giants like Brontosaurs. There are soaring Pterosaurs and swimming Pleiosaurs. Big and small there are a whole world of dinosaurs.

Early birds (Archaeopteryx…ar-kee-OP-ter-iks) appear, too, swooping through the sky, but still no fossils of people.

And what about Rocky? When we last saw him he was a
Sed-i-men-ta-ry (SANDWICH) rock at the bottom of the sea.

But wait! The earth's plates are moving again
and...so is Rocky!
This time Rocky thinks:
Geo-glee, a new adventure for me.
What do YOU think it can possibly be?!

The earth isn't really a solid ball. The crust is more like a giant puzzle where some large and some
smaller puzzle pieces, called plates, are floating like rafts, on the earth's hot molten lava center. They are
always moving but very VERY slowly: only 1 to 2 inches per year. They can move toward each other or
away from each other or next to each other very VERY slowly, over thousands and millions of years.

WHERE do you think Rocky's going to go this time?

ROCKS are made of building blocks called MINERALS. There are thousands of kinds of minerals. Rocks can be made up of different combinations of minerals. Detectives use tests to discover clues. Geologists do too! They test rocks and minerals to find out what they are made of and how they were formed.

CAN YOU FIGURE OUT THESE MYSTERY MINERALS?

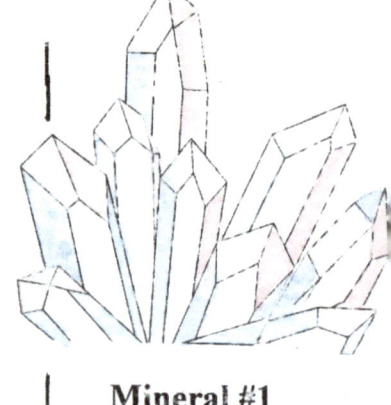

Mineral #1

1) WHAT COLOR ARE your mystery minerals?

TRY Mineral #2
Clues: *gold, shiny, greenish, very hard*

2) HOW SHINY is your mystery mineral...this is called LUSTER?
   It might be: Glassy or Waxy/Greasy or Metallic (shiny) or Dull

3) WHAT COLOR STREAK DID YOUR ROCK MAKE?
   If you SCRATCH your mystery mineral across a clean white tile or "streak plate", it may make a colored STREAK, hard to see if it's white.
   MINERAL #1 doesn't make a streak, or if it does, it's white

4) HOW HARD IS YOUR MINERAL? Depending on what can scratch a mineral, there is a hardness scale of 1 to 10. If your fingernail scratches your mineral, it has a hardness of 2 1/2 or less. If a penny makes a scratch on your mineral, it has a hardness of 3 ½ or less. If a nail makes a scratch on your mineral, it has a hardness of 5 ½ or less. Your mineral #1 is very hard; too hard to be scratched by a nail.

5) Use the chart to solve the mystery. WHICH MINERALS are pictured?

MINERAL IDENTIFICATION CHART

Color	Luster	Streak	Hardness	MINERAL
white, gray	soapy, greasy	none (white)	1	Talc
greenish	silky, waxy	greenish-black	3	Serpentine
brassy gold	shiny	greenish-black	too hard to be	Pyrite
colorless, white	glassy	none (white)	scratched by a nail	Quartz

The puzzle piece that Rocky is part of is pushed down this time. He isn't pushed far enough down as to make him red-hot lava (or magma) like he was before. Never-the-less, under the earth with all the weight of the mountains above him and the heat of the magma below, Rocky is again changed.

Rocky's mineral building blocks are rearranged by the heat and pressure so that now he's a Met-a-mor-phic rock again. That is the name for a CHANGED rock. Like a caterpillar goes through metamorphosis to become a butterfly, Rocky has been changed into a different kind of rock.

Can you find Rocky in his circle of travels?

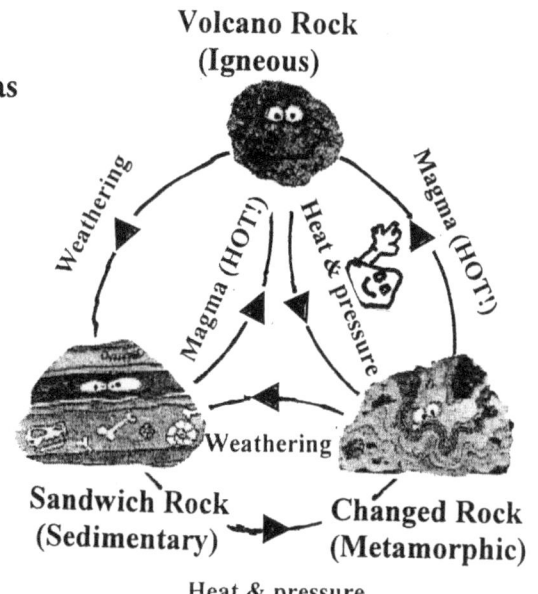

marble   schist   slate

Rocks like Marble, Schist and Slate are metamorphic.

"Hi, there Rocky," says a Sed-i-men-ta-ry rock nearby.

Rocky looks closely.

"Oh, is that you Ned?! You're Sed-i-men-ta-ry now. It's nice, isn't it, though a bit sandy."

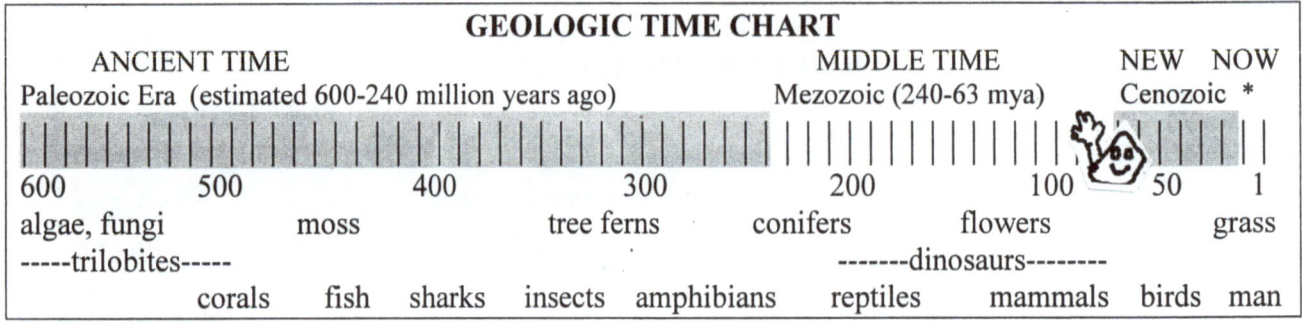

GEOLOGIC TIME CHART

ANCIENT TIME		MIDDLE TIME	NEW NOW
Paleozoic Era (estimated 600-240 million years ago)		Mezozoic (240-63 mya)	Cenozoic *

600	500	400	300	200	100	50	1
algae, fungi	moss	tree ferns	conifers		flowers		grass

-----trilobites-----   -------dinosaurs--------

corals   fish   sharks   insects   amphibians   reptiles   mammals   birds   man

Now it's 80 million years ago ... how long ago is that?...(a long, LOONG time ago!)
Where is it on our chart? And when in the world are the fossils of people going to come along?
Not yet! But, at least, now there are FLOWERS! Can you image a world without flowers?

The first plants to live on land didn't have flowers or seeds. They were mosses and ferns with spores, instead. Next came the conifers which had "real" seeds. And, finally, flowers appear like the ones we know and love today. Some of the first flowers were probably large white ones like Magnolia.

Meanwhile, the first mammals are starting to appear, too. The Eupantotheres (call him Eupy), a small opossum-like creature, ate insects and ran around at night when the dinosaurs were asleep.

16

But, WHAT's happening with Rocky now? Oh no!, those puzzle-piece-plates of the earth are slowly shifting again. "Geo-glee, a new adventure for me. What do YOU think it can possibly be?" This time the plate Rocky is on goes Up and UP. There's our Rocky right on the top of a new mountain...one of our very own mountains.

WHAT MIGHT HAPPEN??!!

1) Inch... by... inch (drum roll, please)two plates move up on each other. Pretend your hands held flat are the plates and try it. One or both are forced to move up.
WHAT might happen?

A lake will be made?
A mountain will be made?
A star will be made?

2) Sometimes the plates move apart. This can form a valley or make a volcano where hot magma that rises through the crack and cools.
WHAT might happen?

A volcano will be made?
A dinosaur will be squished?
A forest will be made?

3) SOMETIMES, two plates move side-to-side and get caught on each other.This causes great tension to build.
WHAT might happen?

A thunderstorm?
A tornado?
An earthquake?

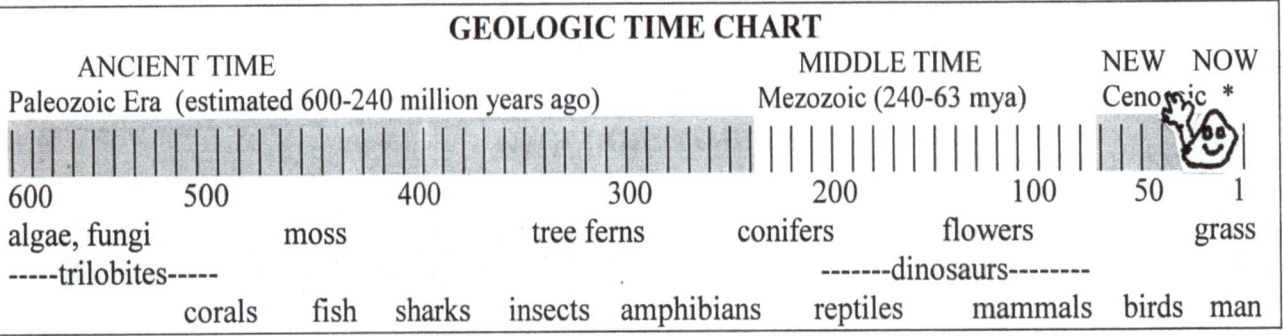

GEOLOGIC TIME CHART

ANCIENT TIME	MIDDLE TIME	NEW NOW
Paleozoic Era (estimated 600-240 million years ago)	Mezozoic (240-63 mya)	Cenozic *

600   500   400   300   200   100   50   1

algae, fungi   moss   tree ferns   conifers   flowers   grass

-----trilobites-----   -------dinosaurs--------

corals   fish   sharks   insects   amphibians   reptiles   mammals   birds   man

Now it's about 15 million years ago. Is that a long time ago? It's still a long, LOOONG time ago. The dinosaurs have disappeared (except the bird-like ones).

When they were gone, MAMMALS started to cover the earth, along with insects and plants and lots of other things. Are we mammals? Yes, but there are still no people on earth. There are, however, tiny camels, three-toed horses and saber-toothed cats screeching through the night in valleys not far from where we live today.

Now it's about 1 million years ago. Where is that on our time chart?

What happens now? The fossils of PEOPLE finally start to appear on earth! Is this a long time ago? Yes, it certainly is. But, looking at our chart, it's not really so very, very long ago, after all, is it?

What else is happening? The weather starts to get colder and colder and colder and the Ice Age begins. Ice is powerful, and as it moves south, its weight and its cold kill many things. Even Rocky on his mountain top feels the ice spreading and creaking slowly by, rubbing and polishing him until he is smooth.

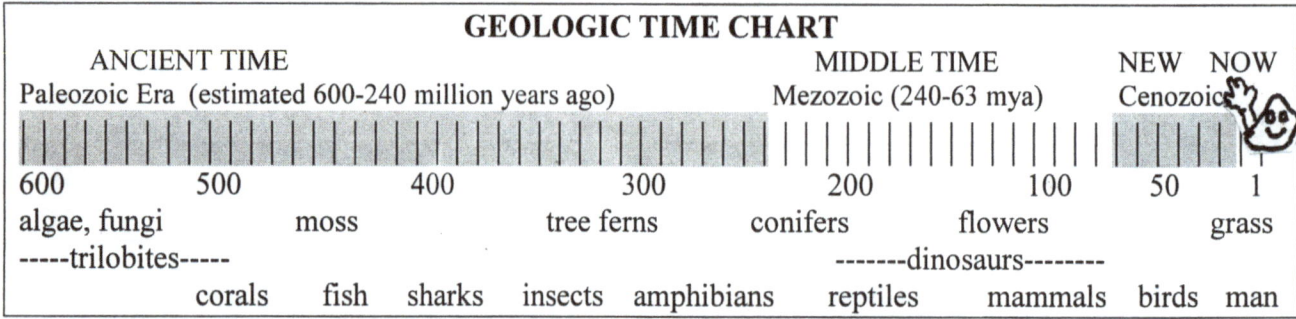

ANCIENT TIME		MIDDLE TIME	NEW	NOW
Paleozoic Era (estimated 600-240 million years ago)		Mezozoic (240-63 mya)	Cenozoic	

600	500	400	300	200	100	50	1
algae, fungi	moss		tree ferns	conifers	flowers		grass
-----trilobites-----				-------dinosaurs--------			
	corals	fish	sharks	insects	amphibians	reptiles	mammals	birds	man

Now it's LAST NIGHT...THAT'S not very long ago at all, is it?! Think of what you did last night. The climate is warmer again, now. The ice has shrunk back to the North and South Poles and mountain tops. But up in the mountains last night it was rainy and cold. Freezing water expanded a crack that caused Rocky to break off his mountain ledge and roll down the mountain side... crash, bang, thunk.

Do you think that's the end of Rocky's story?!

NO WAY! The rock cycle goes on and on and on in a dance that circles, every-changing, never-ending, always amazing.

As Rocky would say,
" Geo-glee, a new adventure for me. What do YOU think it can possibly be?"

20

DISCOVERY PAGES
A Rock of Your Own

Sometimes rocks are made of layers of sand or pebbles (they're called sed-i-men-tary). Sometimes they were formed by a fire-breathing volcano (they are ig-nee-us). Sometimes the crystal building blocks of rocks are mixed up like scrambled eggs and remade by the heat and pressure inside the earth (that's called met-a-more-fic). However a rock is made, it can be nice to have a rock of your very own to rub, hold, carry, or even to talk to. Rocks can be good listeners. Sometimes they might even understand things...like a dog with a wet nose and wagging tail might.

How do you find a rock that is just right for you? Relax and take your time. Don't worry about it...when it's time, you'll find one that's just right for YOU. If you don't find one today, you'll find one another day.

Your parents and your teacher have good ideas that are important to listen to. But when you're looking for a rock, don't let anyone, not even your friends, help to make your decision. This is YOUR rock. Only you know what's right for you.

Look at shapes and colors. A green one might be perfect or a sparkly white one...or sometimes just a smooth grey rock is nice...not showy, just friendly and comfortable. It doesn't have to be beautiful to be the right rock, it just has to make you feel good.

Looking at rocks under water sometimes lets you see colors that are hiding. Other people might not see your rock's beauty...but YOU will! Don't worry what anyone else thinks about your rock. You don't even have to show it to anyone, if you don't want to.

Feel the texture (whether it's rough or smooth). A rough one might be nice to run your fingers over when you're thinking hard, a smooth one might be nice to hold when you're afraid...we're ALL afraid sometimes.

It shouldn't be too large to fit in your hand or too heavy for your pocket. And it shouldn't be too small or it might get lost. Sometimes, though, rocks do get lost and find a new home or, like Rocky, carry on with their journey...that's OK. Your rock would want you to find a new special rock to take its place.

You can start looking for your rock anytime you'd like.

START a ROCK COLLECTION

Label each rock you find with a number.

In a notebook keep track of the date and place you got each rock and what kind of rock you think it is

Keep your rocks safe in an egg carton or box.

ROCK IDENTIFICATION and HOW ROCKS ARE FORMED

IGNEOUS ROCKS: VOLCANO ROCKS, when molten/melted rock cools and crystallizes

What to look for in identifying Igneous rocks:

A rock with crystals,   A rock that does not have bands or layers of colors,

A rock that will scratch a brick,   A rock that is rough outside and filled with bubble holes

Some examples:

OBSIDIAN: black and glassy

PUMICE: tan to cream, VERY light weight with lots of holes

GRANITE: course texture with pink, gray and black speckles; formed underground so it has larger crystals that can twinkle

METAMORPHIC ROCKS: CHANGED ROCKS, their minerals re-crystallized due to heat and pressure without melting.

What to look for:

A layered rock that contains mica,   A rock that breaks into thin leaf-like layers

A rock with layers of dark and light minerals,   A rock with a sugary texture

Some Examples:

SERPENTINE: blue to green, smooth & glassy, soapy-looking

SCHIST: silver flakes, splits into layers easily

SEDIMENTARY ROCKS: SANDWICH ROCKS made from layers of other rocks (any kind) which have weathered to sediment, from plant and animal remains, or from minerals crystallized as water evaporates. Fossils are often found in sedimentary rocks.

What to look for:

A rock made of broken bits and pieces of other rocks,   A rock that shows layers

A rock containing fossils, plant material or shell fragments,   A rock with a dull luster

Some Examples:

SANDSTONE: sand particles layered & compressed: solid colored or banded in yellow, red, tan, orange

CONGLOMERATE: a course texture with lots of pebbles visible

GLOSSARY

KINDS of ROCKS / STEPS in the ROCK CYCLE

IGNEOUS: "Volcano Rock": erupted from a volcano
SEDIMENTARY: "Sandwich Rock": other rocks weathered to sand, minerals &
 organic material pressed together in layers
METAMORPHIC: "Changed Rock": changed under pressure &/or heat

Some PARTS of ROCKS

MINERAL: basic inorganic BUILDING BLOCKS from which rocks are
made; their properties remain standard...they are always the same
QUARTZ: one kind of mineral that is found in many kinds of rocks or by
itself. It is white or colorless and very hard...it will scratch glass.
ROCK: can be made of different combinations of minerals and organic
(once living) material. Rocks can be igneous, metamorphic or sedimentary.
FOSSIL: traces of a plant or animal which has died in an earlier geologic age and
left its print in a rock (usually sedimentary).
TRILOBITE: One kind of fossil. 500 million years ago and for over a hundred million years
TRI-luh-bites (many kinds of them) lived in the seas. Their name means "three sections".
Though they all became extinct at the end of the Paleozoic Era (240 million years ago), they are
ancestors of the horseshoe crab.

Some KINDS of ANIMALS

AMPHIBIAN: animal who lives partly in water and partly on land; they
metamorphose (change) from one form to other as a tadpole becomes a frog.
Amphibians are cold-blooded and usually have smooth skin.
REPTILE: cold-blooded animals who have scaly skin. Reptiles generally lay eggs and
usually live in dry places (although not the crocodile).
MAMMAL: warm-blooded (body temperature is the same all the time) animals who
have babies that are alive and feed them with milk.

Make Some CRYSTALS,
the building blocks of minerals, which are the building blocks of rocks.

Having a parent help you, bring 2 cups of water to a boil in a saucepan.
Stir as you gradually add 4 cups of sugar until it is dissolved.
Pour solution into a glass jar.
Cut a piece of cotton string almost a long as the jar
and tie one end to the middle of a pencil.
Fasten a paper clip to the other end of the string.
Dip the string in the sugar solution,
then straighten it out on wax paper and let it dry overnight.
Now place the pencil across the rim of the jar of sugar-water and suspend
the string so the clip at the bottom is hanging into the solution.
Place the jar where it won't be disturbed or moved; cover with paper
towel.
Check them after several days to a week.
What shape are the crystals?
Try this experiment using salt.
You only need 2 cups of salt for 2 cups of water.
Are the salt crystals different from the sugar crystals?

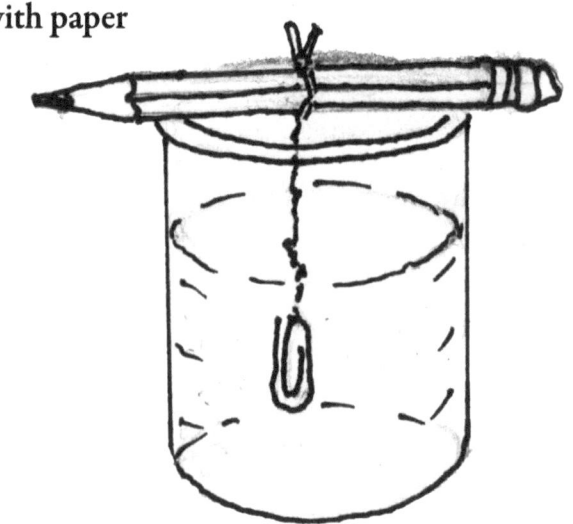

MAKE a Pretend "FOSSIL"

1: Put a 1 inch layer of sand in a shallow container.
2: Put Vaseline/Petroleum Jelly on both sides of the leaf, fern or object you want to make into a "fossil."
3: Lay (don't bury) the objects on the sand.
4: Mix equal amount of water and Plaster of Paris
5: Add just enough water to moisten, thin enough to pour without being soupy
6: Pour/spread the paste evenly over the sand with its "fossil" in a 2 inch layer
7: Put pan in warm place until mixture is hard (½-1 hour)
8: Turn the pan over carefully and clean the sand off to see your new "fossil".

Just a Simple Thing

I'm teased about picking up gravel wherever I go.
It just boggles my mind, though,
what's there at our feet.
A handful of pebbles polished in the tumbler of time:
just a simple thing...but what stories it could tell.

Unfathonable history, our history,
written layer by layer
on a parchment of sandstone.
Trilobites, ammonites,
not so different than we,
   except that we are so very young
   and they, from a world long gone...
      lost in the alluvium of time.

Striped rocks, speckled rocks, green rocks, and pink
   all riding an unending moebius roller coaster,
Lives shaped by the rigid rules of chemistry
and fate.

So many kinds with so many stories.
How did they get here? Where are they going?
Millions of years of geologic history
here in my hand...

Just a simple thing.

- Peggy Price

www.ingramcontent.com/pod-product-compliance
Lightning Source LLC
Chambersburg PA
CBHW050415180526
45159CB00005B/2279